by Adrian Harrison

INTRODUCTION TO LIMITS

January 2020

CONTENTS

LIMIT

Definition

If f(x) approaches to L as x approaches to a, the limit of f(x) is L and it is shown by.

$$\lim_{x \to a} f(x) = L$$

PROPERTIES

1. $\lim\limits_{x \to a} f(x) = f(a)$

2. $\lim\limits_{x \to x_0} k = k \ (k \in R)$

3. $\lim\limits_{x \to x_0} (f(x) \mp g(x)) = \lim\limits_{x \to x_0} f(x) \mp \lim\limits_{x \to x_0} g(x)$

4. $\lim\limits_{x \to x_0} (f(x) \cdot g(x)) = \lim\limits_{x \to x_0} f(x) \cdot \lim\limits_{x \to x_0} g(x)$

5. $k \in R, \ \lim\limits_{x \to x_0} [k \cdot f(x)] = k \cdot \lim\limits_{x \to x_0} f(x)$

6. $\lim\limits_{x \to x_0} \dfrac{f(x)}{g(x)} = \dfrac{\lim\limits_{0 \to x_0} f(x)}{\lim\limits_{x \to x_0} g(x)} \left(\lim\limits_{x \to x_0} g(x) \neq 0 \right)$

7. $\lim\limits_{x \to x_0} [(f(x))^n] = \left[\lim\limits_{x \to x_0} f(x) \right]^n$

8. $\lim\limits_{x \to x_0} \sqrt[n]{f(x)} = \sqrt[n]{\lim\limits_{x \to x_0} f(x)}$

1

$(f(x) > 0 \text{ and } n \text{ is an odd natural number})$

9. $\lim\limits_{x \to x_0} \left[C^{f(x)} \right] = C^{\lim\limits_{x \to x_0} f(x)} \quad C \in R$

10. $\lim\limits_{x \to x_0} \left[\log_a f(x) \right] = \log_a \left[\lim\limits_{x \to x_0} f(x) \right]$

Example:

$f(x) = x^3 + 2x^2 - 3x + 2 \Rightarrow \lim\limits_{x \to 2} f(x) = ?$

Solution:

$\lim\limits_{x \to 2} f(x) = f(2) = 2^3 + 2 \cdot 2^2 - 3 \cdot 2 + 2$

$\lim\limits_{x \to 2} f(x) = f(2) = 8 + 8 - 6 + 2 = 12$

Example:

$f(x) = \dfrac{x^3 + x + 3}{x^2 + 2} \Rightarrow \lim\limits_{x \to 3} f(x) = ?$

Solution:

$\lim\limits_{x \to 3} f(x) = \lim\limits_{x \to 3} \dfrac{3^3 + 3 + 3}{3^2 + 2} = \lim\limits_{x \to 3} \dfrac{33}{11} = 3$

2

UNCERTAINITIES

$$\frac{\infty}{\infty}, \infty - \infty, 0 \cdot \infty, 0^0, \infty^0, 1^\infty$$

(Such types of expressions are called uncertainities).

$$a)\frac{0}{0} \rightarrow \lim_{x \to a} \frac{f(x)}{g(x)} = \frac{0}{0}$$

To solve such types of limits, factorise $f(x)$ and $g(x)$,

then simplify some terms.

Example:

$$f(x) = \frac{x^2 - 1}{x^3 - 1} \Rightarrow \lim_{x \to 1} f(x) = ?$$

Solution:

$$\lim_{x \to 1} f(x) = \lim_{x \to 1} \frac{x^2 - 1}{x^3 - 1} = \frac{\lim_{x \to 1}(x^2 - 1)}{\lim_{x \to 1}(x^3 - 1)} = \frac{1-1}{1-1} = \frac{0}{0}$$

$$\lim_{x \to 1} f(x) = \lim_{x \to 1} \frac{x^2 - 1}{x^3 - 1} = \lim_{x \to 1} \frac{(x-1)(x+1)}{(x-1)(x^2 + x + 1)}$$

$$= \lim_{x \to 1} \frac{x+1}{x^2+x+1} = \frac{1+1}{1+1+1} = \frac{2}{3}$$

Example:

$$f(x) = \frac{\sqrt{x}-2}{x^3-64} \Rightarrow \lim_{x \to 4} f(x) = ? \boxed{10}$$

Solution:

$$\lim_{x \to 4} f(x) = \lim_{x \to 4} \frac{\sqrt{x}-2}{x^3-64} = \frac{\sqrt{4}-2}{4^3-64} = \frac{2-2}{64-64} = \frac{0}{0}$$

$$\lim_{x \to 4} f(x) = \lim_{x \to 4} \frac{(\sqrt{x}-2)(\sqrt{x}+2)}{(x-4)(x^2+4x+16)(\sqrt{x}+2)}$$

$$= \lim_{x \to 4} \frac{x-4}{(x-4)(x^2+4x+16)(\sqrt{x}+2)}$$

$$= \lim_{x \to 4} \frac{1}{(x^2+4x+16)(\sqrt{x}+2)}$$

$$= \frac{1}{(16+16+16)\cdot(2+2)} = \frac{1}{48\cdot4} = \frac{1}{192}$$

$$b) \frac{\infty}{\infty} \to \lim_{x \to \pm\infty} \frac{a_n x^n + a_{n-1} x^{n-1} + \dots a_1 x_1 + a_0}{b_m x^m + b_{m-1} x^{m-1} + \dots + b_1 x + b_0} \boxed{10}$$

$$n > m \Rightarrow limit = \pm\infty$$

$$n = m \Rightarrow limit = \frac{a_n}{b_m}$$

4

$$n < m \Rightarrow limit = 0$$

Example:

$$f(x) = \frac{3x^2 + 4x - 5}{6x^2 + x + 3} \Rightarrow \lim_{x \to \infty} f(x) = ?$$

Solution:

$$\lim_{x \to \infty} f(x) = \lim_{x \to \infty} \frac{3x^2 + 4x - 5}{6x^2 + x + 3} = \frac{\infty + \infty - 5\square}{\infty + \infty + 3} = \frac{\infty}{\infty}$$

$$\lim_{x \to \infty} f(x) = \lim_{x \to \infty} \frac{x^2\left(3 + \dfrac{4}{x} - \dfrac{5}{x^2}\right)}{x^2\left(6 + \dfrac{1}{x} + \dfrac{3}{x^2}\right)} \square$$

$$\lim_{x \to \infty} \frac{3 + 0 - 0}{6 + 0 + 0} = \frac{3}{6} = \frac{1}{2}\square$$

c) $\infty - \infty$ and $0 \cdot \infty$

To solve these types of limits, $\infty - \infty$ and $0 \cdot \infty$

uncertainities have to be converted into $\dfrac{\infty}{\infty}$ or $\dfrac{0}{0}$ types.

Then apply the rule expressed in a and b.

Example:

$$f(x) = 2\sqrt{x^2 + 1} - \sqrt{4x^2 + 2x + 3} \Rightarrow \lim_{x \to \infty} f(x) = ?$$

Solution:

$$\lim_{x \to \infty} \frac{\left(2\sqrt{x^2 + 1} - \sqrt{4x^2 + 2x + 3}\right)\left(2\sqrt{x^2 + 1} + \sqrt{4x^2 - 2x + 3}\right)}{2\sqrt{x^2 + 1} + \sqrt{4x^2 + 2x + 3}}$$

$$= \lim_{x \to \infty} \frac{4x^2 + 4 - 4x^2 - 2x - 3}{x\left(2\sqrt{1 + \dfrac{1}{x^2}} + \sqrt{4 + \dfrac{2}{x} + \dfrac{3}{x^2}}\right)} = \lim_{x \to \infty} \frac{x\left(-2 + \dfrac{1}{x}\right)}{x(2 + 2)} = -\frac{1}{2}$$

d) 1^{∞} *Uncertainities*

Example:

$$f(x) = \left(1 + \frac{3x}{x^2 + 2}\right)^{2x} = ?$$

Solution:

$$\lim_{x \to \infty} f(x) = \lim_{x \to \infty} \left(1 + \frac{3}{x + \dfrac{2}{x}}\right)^{2x}$$

$$= \lim_{x \to \infty} (1 + 0)^{\infty} = 1^{\infty}$$

$$u(x) = \frac{3x}{x^2 + 2}$$

$$\lim_{x \to \infty} u(x) = \lim_{x \to \infty} \frac{3x}{x^2 + 2} = 0$$

$$\vartheta(x) = 2x$$

$$\lim_{x \to \infty} \vartheta(x) = \infty$$

$$u(x) \cdot \vartheta(x) = \frac{6x^2}{x^2 + 2}$$

$$\lim_{x \to \infty} [(u(x) \cdot \vartheta(x))] = \lim_{x \to \infty} \frac{6x^2}{x^2 + 2} = 6$$

$$\lim_{x \to \infty} f(x) = \lim_{x \to \infty} \left(1 + \frac{3x}{x^2 + 2} \right)^{2x} = e^6$$

Example:

$$f(x) = \left(e^{2/x} + \frac{2}{x} \right)^x \Rightarrow \lim_{x \to \infty} f(x) = ?$$

Solution:

$$\lim_{x \to \infty} f(x) = \lim_{x \to \infty} \left(e^{2/x} + \frac{2}{x} \right)^x$$

$$= \lim_{x \to \infty} \left(e^{2/\infty} + \frac{2}{\infty} \right)^\infty = \left(e^0 + 0 \right)^\infty = 1^\infty$$

$$y = \left(e^{2/x} + \frac{2}{x} \right)^x \Rightarrow \ln y = x \cdot \ln \left(e^{2/x} + \frac{2}{x} \right)$$

$$\lim_{x\to\infty} \frac{\ln\left(e^{2/x} + \dfrac{2}{x}\right)}{\dfrac{1}{x}} = \lim_{x\to\infty} \frac{-\dfrac{2}{x^2}e^{2/x} - \dfrac{2}{x^2}}{-\dfrac{1}{x^2}} = 2(e^0 + 1) = 4$$

$$\ln y = 4 \Rightarrow \lim_{x\to\infty} y = e^4$$

LIMITS OF TRIGONOMETRIC FUNCTIONS

1. $\lim\limits_{x\to a} f(x) = f(a) = \sin a$

2. $\lim\limits_{x\to a} g(x) = g(a) = \cos a$

2. $\lim\limits_{x\to 0} \dfrac{\sin x}{x} = 1$

3. $\lim\limits_{x\to 0} \dfrac{\sin ax}{bx} = \dfrac{a}{b}$

4. $\lim\limits_{x\to 0} \dfrac{\tan x}{x} = 1$

Example:

$$f(x) = \frac{\sin 5x}{x} = \ ?$$

Solution:

$$\lim_{x\to 0} f(x) = \lim_{x\to 0} \frac{5 \cdot \sin 5x}{5x}$$

8

$$= 5 \lim_{x \to 0} \frac{\sin 5x}{5x} \boxed{}$$

$$\lim_{x \to 0} f(x) = 5 \lim_{u \to 0} \frac{\sin u}{u} = 5 \cdot 1 = 5 \boxed{}$$

Example:

$$\lim_{x \to 0} \frac{1 - \cos 2x}{4x^2} = ?$$

Solution:

$$\lim_{x \to 0} \frac{1 - \cos 2x}{4x^2} = \lim_{x \to 0} \frac{1 - \cos 0}{4 \cdot 0} = \frac{1 - 1}{0} = \frac{0}{0} \boxed{}\boxed{}$$

$$\lim_{x \to 0} \frac{1 - \cos 2x}{4x^2} = \lim_{x \to 0} \frac{1 - (1 - 2\sin^2 x)}{4 \cdot x^2} \boxed{}\boxed{}$$

$$\lim_{x \to 0} \frac{1 - \cos 2x}{4x^2} = \lim_{x \to 0} \frac{2\sin^2 x}{4x^2} \boxed{}\boxed{}$$

$$= \frac{1}{2} \lim_{x \to 0} \left(\frac{\sin x}{x}\right)^2 = \frac{1}{2} \boxed{}$$

Example:

$$f(x) = \frac{\sin 4x \cdot \tan 2x}{1 - \cos 2x} \Rightarrow \lim_{x \to 0} f(x) = ? \boxed{}$$

Solution:

$$\lim_{x \to 0} f(x) = \lim_{x \to 0} \frac{\sin 4x \cdot \tan 2x}{1 - \cos 2x}$$

$$= \lim_{x \to 0} \frac{\sin 0 \cdot \tan 0}{1 - \cos 0}$$

$$= \frac{0 \cdot 0}{1 - 1} = \frac{0}{0}$$

$$\lim_{x \to 0} f(x) = \lim_{x \to 0} \frac{2 \cdot \sin 2x \cdot \cos 2x \cdot \dfrac{\sin 2x}{\cos 2x}}{1 - (1 - 2\sin^2 x)}$$

$$= \lim_{x \to 0} \frac{2\sin^2 2x}{2\sin^2 x}$$

$$= \lim_{x \to 0} \frac{2 \cdot 4\sin^2 x \cdot \cos^2 x}{2\sin^2 x}$$

$$= \lim_{x \to 0} 4\cos^2 x = 4 \cdot \cos 0$$

$$= 4 \cdot 1 = 4$$

Example:

$$\lim_{x \to \pi} \frac{\cos 2x - 1}{\sin^2 x} = ?$$

Solution:

$$\lim_{x \to \pi} \frac{\cos 2x - 1}{\sin^2 x} = \lim_{x \to \pi} \frac{\cos 2\pi - 1}{\sin^2 \pi}$$

10

$$= \frac{1-1}{0} = \frac{0}{0}$$

$$\lim_{x \to \pi} \frac{\cos 2x - 1}{\sin^2 x} = \lim_{x \to \pi} \frac{1 - 2\sin^2 x - 1}{\sin^2 x}$$

$$= \lim_{x \to \pi} \frac{-2\sin^2 x}{\sin^2 x} = -2$$

TEST WITH SOLUTIONS

1. $\lim\limits_{x \to -1} \dfrac{x^3 + 5x^2 - 2x}{3x^4 - 2x^3 - 4x^2 + 1} = ?$

A) 1   B) 2   C) 3   D) 4   E) 5

Solution:

$$\lim_{x \to (-1)} \frac{x^3 + 5x^2 - 2x}{3x^4 - 2x^3 - 4x^2 + 1}$$

$$= \frac{(-1)^3 + 5\cdot(-1)^2 - 2\cdot(-1)}{3\cdot(-1)^4 - 2\cdot(-1)^3 - 4\cdot(-1)^2 + 1}$$

$$= \frac{-1 + 5 + 2}{3 + 2 - 4 + 1}$$

$$= \frac{6}{2}$$

$$= 3$$

Correct Answer : C

2. $\lim\limits_{x \to 3} \dfrac{x^2 - 9}{x - 3} = ?$

A) 0   B) 3   C) 5   D) 6   E) 9

Solution:

$$\lim\limits_{x \to 3} \dfrac{3^2 - 9}{3 - 3} = \dfrac{0}{0}$$

$$\lim\limits_{x \to 3} \dfrac{x^2 - 9}{x - 3} = \lim\limits_{x \to 3} \dfrac{(x - 3)(x + 3)}{x - 3}$$

$$= 3 + 3$$

$$= 6$$

Correct Answer : D

3. $\lim\limits_{x \to -3} \dfrac{x^3 + 8}{x + 2} = ?$

A) 15   B) 19   C) 21   D) 27   E) 30

Solution:

$$\lim\limits_{x \to -3} \dfrac{x^3 + 8}{x + 2} = \dfrac{(-3)^3 + 8}{-3 + 2}$$

$$= \dfrac{-27 + 8}{-1}$$

$$= \dfrac{-19}{-1}$$

$$= 19$$

4. $\lim\limits_{x \to 3} \dfrac{\cos x - \sin 3°}{\sin x - \cos 3°} = ?$

A) – 2   B) – 1   C) 0   D) 1   E) 2

Solution:

$$\lim\limits_{x \to 3} \dfrac{\cos x - \sin 3°}{\sin x - \cos 3°} = \dfrac{\cos 3° - \sin 3°}{\sin 3° - \cos 3°}$$

$$= -1$$

5. $\lim\limits_{x \to \frac{\pi}{6}} \dfrac{\cot x - 2\cos x}{\sin x + \tan x} = ?$

A) – 2   B) – 1   C) 0   D) 1   E) 2

Solution:

$$\lim\limits_{x \to \frac{\pi}{6}} \dfrac{\cot x - 2\cos x}{\sin x + \tan x} = \dfrac{\cot \dfrac{\pi}{6} - 2\cos \dfrac{\pi}{6}}{\sin \dfrac{\pi}{6} + \tan \dfrac{\pi}{6}}$$

$$= \dfrac{\sqrt{3} - 2 \cdot \dfrac{\sqrt{3}}{2}}{\dfrac{1}{2} + \dfrac{\sqrt{3}}{3}}$$

$$= \dfrac{\dfrac{\sqrt{3} - \sqrt{3}}{3 + 2\sqrt{3}}}{6}$$

13

$$= 0$$

6. $\lim\limits_{x \to 2} \dfrac{x^3 - 2x - 4}{x^3 - 8} = ?$

A) 1   B) $\dfrac{3}{2}$ C) 2   D) $\dfrac{5}{6}$ E) $\dfrac{6}{5}$

Solution:

$$\lim\limits_{x \to 2} \frac{x^3 - 2x - 4}{x^3 - 8} = \frac{8 - 4 - 4}{8 - 8} = \frac{0}{0}$$

$$\lim\limits_{x \to 2} \frac{x^3 - 2x - 4}{x^3 - 8} = \lim\limits_{x \to 2} \frac{(x - 2)(x^2 + 2x + 2)}{(x - 2)(x^2 + 2x + 4)}$$

$$= \frac{2^2 + 2 \cdot 2 + 2}{2^2 + 2 \cdot 2 + 4}$$

$$= \frac{10}{12}$$

$$= \frac{5}{6}$$

Correct Answer: D

7. $\lim\limits_{x \to 0} \dfrac{\sin 3x}{x} = ?$

A) – 3   B) – 1   C) 0   D) 1   E) 3

Solution:

$$\lim_{x \to 0} \frac{\sin 3x}{x} = 3 \cdot \lim_{(3x) \to 0} \frac{\sin 3x}{3x} = 3 \boxed{?}$$

Correct Answer : E

8. $\lim\limits_{x \to 1} \dfrac{sin(x-1)}{x^2 - 1} = ?$

A)$\dfrac{1}{2}$ B) 1   C)$\dfrac{3}{2}$ D) 2   E)$\dfrac{5}{2}$

Solution:

$$\lim_{x \to 1} \frac{sin\boxed{?}(x-1)}{x^2 - 1} = \frac{\sin 0}{0} = \frac{0}{0} \boxed{?}$$

$$\lim_{x \to 1} \frac{sin\boxed{?}(x-1)}{x^2 - 1} = \lim_{x \to 1} \frac{sin\boxed{?}(x-1)}{(x-1)(x+1)} \boxed{?}\boxed{?}$$

$$= \lim_{x \to 1} \frac{sin\boxed{?}(x-1)}{x-1} \cdot \lim_{x \to 1} \frac{1}{x+1} \boxed{?}\boxed{?}$$

$x - 1 = t$

$$x \to 1 \Rightarrow t \to 0 \quad = \lim_{t \to 0} \frac{\sin t}{t} \cdot \lim_{t \to 0} \frac{1}{t+2} \boxed{?}\boxed{?}$$

$$= 1 \cdot \frac{1}{2} = \frac{1}{2}$$

Correct Answer: A

9. $\lim\limits_{x \to 1} \dfrac{\sqrt{x}-1}{x-1} = ?$

A) 1   B)$\dfrac{1}{2}$ C)$\dfrac{3}{2}$ D) 4   E)$\dfrac{5}{2}$

15

Solution:

$$\lim_{x \to 1} \frac{\sqrt{x} - 1}{x - 1} = \frac{\sqrt{1} - 1}{1 - 1} = \frac{0}{0}$$

$$\lim_{x \to 1} \frac{\sqrt{x} - 1}{x - 1} = \lim_{x \to 1} \frac{\sqrt{x} - 1}{(\sqrt{x} - 1)(\sqrt{x} + 1)}$$

$$= \lim_{x \to 1} \frac{1}{\sqrt{x} + 1}$$

$$= \frac{1}{1 + 1} = \frac{1}{2}$$

Correct Answer: B

10. $\displaystyle \lim_{x \to a} \frac{x^3 - a^3}{x - a} = \ ?$

A) $3a$ B) a^2 C) $3a^2$ D) $5a^2$ E) $6a^2$

Solution:

$$\lim_{x \to a} \frac{x^3 - a^3}{x - a} = \frac{a^3 - a^3}{a - a} = \frac{0}{0}$$

$$\lim_{x \to a} \frac{x^3 - a^3}{x - a} = \lim_{x \to a} \frac{(x - a)(x^2 + ax + a^2)}{x - a}$$

$$= \lim_{x \to a} (x^2 + ax + a)$$

$$= a^2 + a \cdot a + a^2$$

$$= 3a^2$$

Correct Answer: C

16

11. $\lim\limits_{x \to 0} \dfrac{\sin 2x}{\sin 5x} = ?$

A) $-\dfrac{5}{2}$ B) $-\dfrac{2}{5}$ C) $\dfrac{1}{5}$ D) $\dfrac{2}{5}$ E) $\dfrac{5}{2}$

Solution:

$$\lim\limits_{x \to 0} \frac{\sin 2x}{\sin 5x} = \lim\limits_{x \to 0} \frac{2x \cdot \dfrac{\sin 2x}{2x}}{5x \cdot \dfrac{\sin 5x}{5x}}$$

$$= \lim\limits_{x \to 0} \frac{2x}{5x} \cdot \frac{\lim\limits_{2x \to 0} \dfrac{\sin 2x}{2x}}{\lim\limits_{5x \to 0} \dfrac{\sin 5x}{5x}}$$

$$= \frac{2}{5} \cdot \frac{1}{1}$$

$$= \frac{2}{5}$$

Correct Answer: D

12. $\lim\limits_{x \to 2} \dfrac{\sin(x^2 - 4)}{x - 2} = ?$

A) -4 B) -2 C) 0 D) 2 E) 4

Solution:

$$\lim\limits_{x \to 2} \frac{\sin(x^2 - 4)}{x - 2} = \frac{0}{0}$$

$$\lim_{x \to 2} \frac{(x+2) \sin(x^2-4)}{x^2-4} =$$

$$x^2 - 4 = t \Rightarrow \lim_{x \to 2} (x^2 - t) = 0$$

$$= \lim_{x \to 2} (x+2) \cdot \lim_{t \to 0} \frac{\sin t}{t}$$

$$= 4 \cdot 1 = 4$$

Correct Answer: E

13. $\lim_{x \to \infty} 5^{2/x} = ?$

A) – 2   B) – 1   C) 0   D) 1   E) 2

Solution:

$$\lim_{x \to \infty} 5^{2/x} = 5^{2/\infty}$$

$$= 5^0$$

$$= 1$$

Correct Answer: D

14. $\lim_{x \to 2} \frac{x^2 + x - 6}{x^2 - 4} = ?$

A) $\frac{5}{4}$ B) 1   C) $\frac{5}{6}$ D) $\frac{5}{8}$ E) $\frac{1}{2}$

Solution:

18

$$\lim_{x \to 2} \frac{x^2 + x - 6}{x^2 - 4} = \frac{4 + 2 - 6}{4 - 4} = \frac{0}{0}$$

$$\lim_{x \to 2} \frac{x^2 + x - 6}{x^2 - 4} = \lim_{x \to 2} \frac{(x + 3)(x - 2)}{(x - 2)(x + 2)} \; \boxed{\text{\tiny JO}}$$

$$= \frac{2 + 3}{2 + 2}$$

$$= \frac{5}{4}$$

Correct Answer : A

15. $\lim\limits_{x \to 2} \dfrac{\sqrt{x + 2} - 2}{x - 2} = ?$

A) $\dfrac{1}{8}$ B) $\dfrac{1}{4}$ C) $\dfrac{1}{2}$ D) 1   E) 2

Solution:

$$\lim_{x \to 2} \frac{\sqrt{x + 2} - 2}{x - 2} = \frac{\sqrt{4} - 2}{2 - 1} = \frac{0}{0} \boxed{\text{\tiny JO JO}}$$

$$= \lim_{x \to 2} \frac{\sqrt{x + 2} - 2}{x - 2}$$

$$= \lim_{x \to 2} \frac{(\sqrt{x + 2} - 2)(\sqrt{x + 2} + 2)}{(x - 2)(\sqrt{x + 2} + 2)}$$

$$= \lim_{x \to 2} \frac{x - 2}{(x - 2)(\sqrt{x + 2} + 2)}$$

$$= \frac{1}{\sqrt{2 + 2} + 2}$$

19

$$= \frac{1}{4}$$

Correct Answer: B

16. $\lim\limits_{x \to \infty} \dfrac{x^3 - 3}{3x^3 + 2x + 1} = ?$

A) $-\dfrac{1}{3}$ B) $-\dfrac{1}{2}$ C) $\dfrac{1}{3}$ D) $\dfrac{1}{2}$ E) 1

Solution:

$$\lim_{x \to \infty} \frac{x^3 - 3}{3x^3 + 2x + 1} = \lim_{x \to \infty} \frac{x^3 \cdot \left(1 - \dfrac{2}{x^2} + \dfrac{5}{x^3}\right)}{x^3 \cdot \left(3 + \dfrac{1}{x^2} - \dfrac{2}{x^3}\right)}$$

$$= \frac{1 - \dfrac{2}{\infty^2} + \dfrac{5}{\infty^3}}{3 + \dfrac{1}{\infty^2} + \dfrac{2}{\infty^3}}$$

$$= \frac{1 - 0 + 0}{3 + 0 - 0}$$

$$= \frac{1}{3}$$

Correct Answer: C

17. $\lim\limits_{x \to \frac{\pi}{3}} \dfrac{3x - \pi}{\cos\dfrac{9x}{2}} = ?$

A) $\dfrac{1}{6}$ B) $\dfrac{1}{3}$ C) $\dfrac{2}{3}$ D) 1   E) $\dfrac{4}{3}$

Solution:

$$\lim_{x \to \frac{\pi}{3}} \frac{3x - \pi}{\cos \frac{9x}{2}} = \frac{0}{0} \left(\lim_{x \to \frac{\pi}{3}} \frac{9x - 3\pi}{2} = 0 \right) \text{\tiny[10]} \text{\tiny[10]}$$

$$\lim_{x \to \frac{\pi}{3}} \frac{3x - \pi}{\cos \frac{9x}{2}} = \frac{3x - \pi}{-\sin\left(\dfrac{3\pi}{2} - \dfrac{9\pi}{2}\right)}$$

$$= \lim_{x \to \frac{\pi}{3}} \frac{\dfrac{9x - 3\pi}{2}}{\dfrac{3}{2}\sin\left(\dfrac{9x - 3\pi}{2}\right)} \text{\tiny[10]}$$

$$= \frac{1}{\dfrac{3}{2}} \cdot 1$$

$$= \frac{2}{3}$$

Correct Answer: C

18. $\lim\limits_{x \to \infty} \dfrac{x^2 - 3}{x^3 + 2x + 1} = ?$

A) 3   B) 2   C) 1   D) 0   E) – 1

Solution:

$$\lim_{x \to \infty} \frac{x^2 - 3}{x^3 + 2x + 1} = \lim_{x \to \infty} \frac{x^2 \cdot \left(1 - \dfrac{3}{x^2}\right)}{x^2 \cdot \left(x + \dfrac{2}{x} + \dfrac{1}{x^2}\right)}$$

$$= \frac{1 - \dfrac{3}{\infty^2}}{\infty + \dfrac{2}{\infty} + \dfrac{1}{\infty^2}}$$

$$= \frac{1 - 0}{\infty + 0 + 0}$$

$$= \frac{1}{\infty}$$

$$= 0$$

Correct Answer: D

19. $\lim\limits_{x \to \infty} \dfrac{x^5 - x^4 + 1}{x^3 + 2x} = ?$

A) $-\infty$ B) $-\dfrac{5}{3}$ C) 0   D) $\dfrac{5}{3}$ E) ∞

Solution:

$$\lim_{x \to \infty} \frac{x^5 - x^4 + 1}{x^3 + 2x} = \lim_{x \to \infty} \frac{x^3 \cdot \left(x^2 - x + \dfrac{1}{x^3}\right)}{x^3 \cdot \left(1 + \dfrac{2}{x^2}\right)}$$

22

$$= \frac{\infty^2 - \infty + \dfrac{1}{\infty^3}}{1 + \dfrac{2}{\infty^2}}$$

$$= \infty$$

Correct Answer: E

20. $\lim\limits_{x \to -1} \dfrac{\sin \pi x}{x + 1} = ?$

A) $-\pi$ B) $-\dfrac{\pi}{2}$ C) 0   D) $\dfrac{\pi}{2}$ E) π

Solution:

$$\lim\limits_{x \to -1} \frac{\sin(\pi x)}{x + 1} = \frac{0}{0}$$

$$x + 1 = t$$

$$x = t - 1$$

$$x \to -1 \implies t \to 0$$

$$\lim\limits_{t \to 0} \frac{-\sin(\pi - \pi t)}{t} = \lim\limits_{t \to 0} \frac{-\pi \sin \pi t}{\pi t} = -\pi$$

Correct Answer: A

21. $\lim\limits_{x \to \pi} \dfrac{1 + \cos x}{1 - \sin\dfrac{x}{2}} = ?$

A) 2   B) 3   C) 4   D) $\dfrac{3}{4}$ E) $\dfrac{9}{2}$

Solution:

$$\lim_{x \to \pi} \frac{1 + \cos x}{1 - \sin \dfrac{x}{2}} = \frac{1 + \cos \pi}{1 - \sin \dfrac{\pi}{2}} = \frac{0}{0}$$

$$\lim_{x \to \pi} \frac{1 + \cos x}{1 - \sin \dfrac{x}{2}} = \lim_{x \to \pi} \frac{-\sin x}{-\dfrac{1}{2}\cos \dfrac{x}{2}}$$

$$\lim_{x \to \pi} \frac{-\cos x}{\dfrac{1}{4}\sin \dfrac{x}{2}} = \frac{-\cos \pi}{\dfrac{1}{4}\sin \dfrac{\pi}{2}}$$

$$= \frac{-(-1)}{\dfrac{1}{4}\cdot 1} = 4$$

Correct Answer: C

QUESTIONS

1. $\lim\limits_{x \to 3} \dfrac{x^2 - 6x + 9}{x^2 - 8x + 15} = \; ?$

A) $-\dfrac{9}{13}$ B) 0   C) 1   D) $\dfrac{9}{13}$ E) $\dfrac{3}{5}$

Solution:

$$\lim_{x \to 3} \frac{x^2 - 6x + 9}{x^2 - 8x + 15} = \lim_{x \to 3} \frac{(x - 3)(x - 3)}{(x - 3)(x - 5)} = \frac{3 - 3}{3 - 5} = 0$$

24

$$2.\lim_{x\to a}\frac{x^3-a^3}{x^2-a^2}= \ ?$$

A)$\dfrac{3a}{2}$ B) a C)$\dfrac{3}{2}$ D) 1   E) 0

Solution:

$$\lim_{x\to a}\frac{x^3-a^3}{x^2-a^2}=\lim_{x\to a}\frac{(x-a)\cdot(x^2+ax+a^2)}{(x-a)\cdot(x+a)}$$

$$=\frac{a^2+a^2+a^2}{a+a}=\frac{3a^2}{2a}=\frac{3a}{2}$$

Correct Answer: A

$$3.\lim_{x\to0}\frac{x^2+2\cdot sin\,x}{\frac{1}{2}\cdot(e^x-e^{x^2})}= \ ?$$

A) 2   B) 3   C) 4   D)$\dfrac{3}{4}$ E)$\dfrac{5}{2}$

Solution:

$$\lim_{x\to0}\frac{x^2+2\cdot sin\,x}{\frac{1}{2}\cdot(e^x-e^{x^2})}=\frac{0}{0}$$

$$\lim_{x \to 0} \frac{2x + 2 \cdot \cos x}{\frac{1}{2} \left(e^x - e^{x^2} \cdot 2x \right)} = \frac{2}{\frac{1}{2}} = 4$$

4. $\lim\limits_{x \to \frac{\pi}{2}} \dfrac{\pi - 2x}{\sin 8x} = \, ?$

A) $-\dfrac{1}{2}$ B) $-\dfrac{1}{4}$ C) 0   D) 1   E) $\dfrac{1}{2}$

Solution:

$$\lim_{x \to \frac{\pi}{2}} \frac{\pi - 2x}{\sin 8x} = \frac{0}{0}$$

$$\Rightarrow \lim_{x \to \frac{\pi}{2}} \frac{\pi - 2x}{\sin 8x} = \lim_{x \to \frac{\pi}{2}} \frac{-2}{\cos 8x \cdot 8}$$

$$= \frac{-2}{(\cos 4\pi) \cdot 8} = -\frac{2}{8} = -\frac{1}{4}$$

5. $\lim\limits_{x \to -1} \dfrac{x^{15} + 1}{x^6 - 1} = \, ?$ 🔟

A) -8 B) 3   C) $\dfrac{5}{2}$ D) $-\dfrac{3}{2}$ D) $-\dfrac{5}{2}$

Solution:

$$\lim_{x \to -1} \frac{x^{15} + 1}{x^6 - 1} = \frac{(-1)^{15} + 1}{(-1)^6 - 1} = \frac{0}{0}$$

$$= \lim_{x \to -1} \frac{15x^{14}}{6x^5} = \frac{15 \cdot (-1)^{14}}{6 \cdot (-1)^5} = -\frac{5}{2}$$

Correct Answer: E

6. $\lim\limits_{x \to -1} \dfrac{x^{36} - 1}{2x^9 + 2} = ?$

A) $-\dfrac{1}{2}$ B) -4 C) -2 D) 2 E) 4

Solution:

$$\lim_{x \to -1} \frac{x^{36} - 1}{2x^9 + 2} = \frac{1 - 1}{-2 + 2} = \frac{0}{0}$$

$$\lim_{x \to -1} \frac{x^{36} - 1}{2x^9 + 2} = \lim_{x \to -1} \frac{(x^{18})^2 - 1^2}{2 \cdot (x^9 + 1)}$$

$$= \lim_{x \to -1} \frac{(x^{18} - 1)(x^{18} + 1)}{2 \cdot (x^9 + 1)}$$

$$= \lim_{x \to -1} \frac{(x^9 - 1)(x^9 + 1)(x^{18} + 1)}{2 \cdot (x^9 + 1)}$$

$$= -2$$

Correct Answer: C

27

7. $\lim\limits_{x \to a} \dfrac{x^2 - a^2}{x^2 - x - ax + a} = ?$ 🔟

A) $2a^2$ B) $2a$ C) a D) $\dfrac{3}{2}a$ E) $\dfrac{2a}{a-1}$

Solution:

$\lim\limits_{x \to a} \dfrac{(x-a)(x+a)}{(x-a)(x-1)} = \dfrac{2a}{a-1}$ 🔟

Correct Answer: E

8. $\lim\limits_{x \to 0} \left(\dfrac{\tan^2 3x}{25x^2} \right) = ?$

A) $\dfrac{3}{25}$ B) $\dfrac{6}{25}$ C) $\dfrac{9}{25}$ D) 6   E) 9

Solution:

$\lim\limits_{x \to 0} \left[\dfrac{\tan 3x}{3x} \cdot \dfrac{\tan 3x}{3x} \cdot \dfrac{9}{25} \right]$ 🔟

$= \dfrac{9}{25} \lim\limits_{x \to 0} \dfrac{\tan 3x}{3x} \cdot \lim\limits_{x \to 0} \dfrac{\tan 3x}{3x} = \dfrac{9}{25}$

Correct Answer: C

9. $\lim\limits_{x \to -\infty} (5^{3/x} + 4^x + 3) = ?$

A) 0   B) 1   C) 2   D) 3   E) 4

28

Solution:

$$\lim_{x \to -\infty} \left(5^{3/x} + 4^x + 3\right) = 5^{3/-\infty} + 4^{-\infty} + 3$$

$$= 5^0 + \frac{1}{4^\infty} + 3 = 4 + \frac{1}{\infty} = 4$$

Correct Answer: E

10. $\lim\limits_{x \to \frac{1}{3}} \left(\dfrac{x^3 - \dfrac{1}{27}}{x^2 - \dfrac{1}{9}}\right) = ?$

A)$\dfrac{1}{2}$ B)$\dfrac{3}{2}$ C)$\dfrac{5}{2}$ D) 2   E) 3

Solution:

$$\lim_{x \to \frac{1}{3}} \frac{\left(x - \dfrac{1}{3}\right)\left(x^2 + \dfrac{x}{3} + \dfrac{1}{9}\right)}{\left(x - \dfrac{1}{3}\right)\left(x + \dfrac{1}{3}\right)}$$

$$= \frac{\left(\dfrac{1}{3}\right)^2 + \dfrac{1}{9} + \dfrac{1}{9}}{\dfrac{1}{3} + \dfrac{1}{3}} = \frac{\dfrac{1}{3}}{\dfrac{2}{3}} = \frac{1}{2}$$

Correct Answer: A

11. $\lim\limits_{x\to\infty} \dfrac{x^2 - x + 3}{-x^5 + x^2} = ?$

A) 0   B) 1   C)$\dfrac{1}{3}$ D)$\dfrac{1}{5}$ E) ∞

Solution:

$$\lim_{x\to\infty} \frac{x^2 - x + 3}{-x^5 + x^2} = \frac{\infty - \infty}{-\infty + \infty}$$

$$\lim_{x\to\infty} \frac{x^2 - x + 3}{-x^5 + x^2} = \lim_{x\to\infty} \frac{x^2\left(1 - \dfrac{1}{x} + \dfrac{3}{x^2}\right)}{x^2(1 - x^3)}$$

$$\lim_{x\to\infty} \frac{1 - 0 + 0}{1 - \infty} = \frac{1}{-\infty} = 0$$

Correct Answer: A

12. $\lim\limits_{x\to 1} \dfrac{2x^3 + x^2 - 1}{x^3 + 3} = ?$

A)$\dfrac{1}{2}$ B)$\dfrac{3}{2}$ C) 2   D) 3   E) 4

Solution:

$$\lim_{x\to 1} \frac{2x^3 + x^2 - 1}{x^3 + 3} = \frac{2\cdot 1^3 + 1^2 - 1}{1^3 + 3}$$

$$= \frac{2}{4}$$

$$= \frac{1}{2}$$

Correct Answer: A

13. $\lim\limits_{x\to 0} \dfrac{\pi \, sin\dfrac{\pi x}{6}}{3x \, cos\dfrac{\pi x}{3}} = \, ?$

A) $\dfrac{\pi}{9}$ B) $\dfrac{\pi}{19}$ C) $\dfrac{\pi^2}{18}$ D) 2π E) $3\pi + 1$

Solution:

$$\lim\limits_{x\to 1} \frac{\pi sin\dfrac{\pi x}{6}}{3x cos\dfrac{\pi x}{3}} = \pi \cdot \lim\limits_{x\to 0} \frac{sin\dfrac{\pi x}{6}}{3x} \cdot \lim\limits_{x\to 1} \frac{1}{cos\dfrac{\pi x}{3}}$$

$$= \pi \cdot \frac{\dfrac{\pi}{6}}{3} \cdot 1$$

$$= \frac{\pi^2}{18}$$

Correct Answer: C

31

Chapter

Limit

Test 1

1. $\lim\limits_{x \to \frac{1}{2}} \dfrac{x^3 - \dfrac{1}{8}}{x^2 - \dfrac{1}{4}} = ?$ 🔟

A) $-\dfrac{4}{3}$ B) $-\dfrac{3}{4}$ C) $\dfrac{1}{8}$ D) $\dfrac{1}{2}$ E) $\dfrac{3}{4}$

2. $\lim\limits_{x \to 1} \left(\dfrac{1}{1-x} - \dfrac{3}{x - x^3} \right) = ?$ 🔟

A) $-\infty$ B) -1 C) 0 D) $\dfrac{1}{2}$ E) 1

3. $\lim\limits_{x \to 2} \dfrac{x^2 - 2x + 4}{x^2 - 5x + 6} = ?$ 🔟

A) $\dfrac{2}{3}$ B) 1 C) 4 D) 6 E) ∞

4. $\lim\limits_{x \to 1} \dfrac{x^2 + x - 2}{x - 1} = ?$

A) -1 B) 0 C) 1 D) 2 E) 3

5. $\lim\limits_{x\to 0} \dfrac{x\cdot \sin 2x}{\sin^2 x} = \ ?$

A) $-\infty$ B) 0   C) 1   D) 2   E) 3

6. $\lim\limits_{x\to\infty} \left(\dfrac{5x^2}{1-x^2} + 2^{\frac{1}{x}} \right) =$

A) -4 B) -3 C) 3   D) 4   E) 8

7. $\lim\limits_{x\to\infty} \left(\dfrac{x^3}{x^2+1} - x \right) = \ ?$

A) -1 B) 0   C) $\dfrac{2}{3}$ D) 1   E) ∞

8. $\lim\limits_{x\to\frac{\pi}{4}} \dfrac{\cos x - \sin x}{\cos 2x} = \ ?$

A) $-\dfrac{\sqrt{2}}{2}$ B) -1 C) $\dfrac{\sqrt{2}}{2}$ D) $\dfrac{\sqrt{3}}{2}$ E) 1

9. $\lim\limits_{x\to 1} \dfrac{\sqrt{x}-1}{\sqrt[3]{x}-1} = \ ?$

A) 3   B) 2   C)$\dfrac{3}{2}$ D)$\dfrac{2}{3}$ E)$\dfrac{1}{2}$

10.$\lim\limits_{a \to x} \dfrac{a^6 - x^6}{x^2 - a^2} = ?$

A) $-3x^4$ B) $-3x^2$ C) $-3x$ D) $3a^2$ E) $3a^3$

11.$\lim\limits_{x \to 1} \dfrac{x - \sqrt{x}}{1 - \sqrt{x}} = ?$

A) -2 B) -1 C) 0   D) 1   E) 2

12. $\lim\limits_{x \to 6} \dfrac{5 - \sqrt{4x + 1}}{6 - x} = ?$

A) $-\dfrac{2}{5}$ B) $-\dfrac{3}{10}$ C)$\dfrac{3}{10}$ D)$\dfrac{2}{5}$ E) 1

13.$\lim\limits_{x \to a} \dfrac{\sin(x - a)}{x^2 - a^2} = ?$

A)$\dfrac{1}{4a^2}$ B)$\dfrac{1}{3a}$ C)$\dfrac{1}{2a}$ D) a E) $2a$

14.$\lim\limits_{x \to 0} \dfrac{\sin^2 x}{1 - \cos x} = ?$

A) – 1   B) $-\dfrac{1}{2}$ C) $\dfrac{1}{2}$ D) 1   E) 2

15. $\displaystyle \lim_{x \to -3} \dfrac{x + \sqrt{6 - x}}{2x + 6} = ?$

A) $-\dfrac{5}{12}$ B) $-\dfrac{1}{12}$ C) $\dfrac{1}{5}$ D) $\dfrac{5}{12}$ E) $\dfrac{1}{2}$

16. $\displaystyle \lim_{x \to 1} \dfrac{2x^2 + 5x - 7}{x - 1} = ?$

A) 5   B) 6   C) 7   D) 8   E) 9

17. $\displaystyle \lim_{x \to \infty} \dfrac{(a - 1)x^2 + 2}{(a - 1)x^2 - 5x} = ?$

A) – 2   B) – 1   C) 0   D) 1   E) 2

18. $\displaystyle \lim_{x \to 2} \dfrac{5x^2 - 3x}{x} = ?$

A) 7   B) $\dfrac{17}{4}$ C) $\dfrac{15}{4}$ D) 2   E) $\dfrac{13}{12}$

19. $\lim\limits_{x\to\frac{\pi}{2}} \dfrac{\sin x - 1}{\cos 2x + 1} = ?$ 🔟

A) 1 　　 B)$\dfrac{1}{2}$ 　　 C) 0 　　 D) $-\dfrac{1}{4}$ 　　 E) -1

20. $\lim\limits_{x\to 3} \dfrac{4 - \sqrt{a - x}}{x - 3} = m,\ m \in R \Rightarrow m = ?$

A)$\dfrac{1}{2}$ 　　 B)$\dfrac{1}{4}$ 　　 C)$\dfrac{1}{8}$ 　　 D) $-\dfrac{1}{8}$ 　　 E) $-\dfrac{1}{16}$

21. $\lim\limits_{b\to 2} \dfrac{\sin(\pi\cdot b)}{4 - b^2} = ?$ 🔟

A) $-\pi$ 　　 B) $-\dfrac{\pi}{2}$ 　　 C) $-\dfrac{\pi}{4}$ 　　 D)$\dfrac{\pi}{4}$ 　　 E)$\dfrac{\pi}{2}$

22. $\lim\limits_{x\to 0} \dfrac{\ln(1 + 4x)}{\sin 5x} = ?$

A)$\dfrac{4}{5}$ 　　 B) 0 　　 C) $-\dfrac{4}{5}$ 　　 D) 1 　　 E)$\dfrac{5}{4}$

23. $\lim\limits_{x\to 2} \dfrac{\tan(2x - 4)}{x - 2} = ?$

A) -2 　　 B) -1 　　 C) 0 　　 D) 1 　　 E) 2

Chapter   **Limit**

Test 2

1. $\lim\limits_{x \to 1} \dfrac{x^8 - 1}{x - 1} = ?$

A) 1   B) 2   C) 4   D) 8   E) 16

2. $\lim\limits_{x \to 9} \dfrac{2\sqrt{x} - 6}{x - 9} = ?$

A) $\dfrac{1}{2}$ B) $\dfrac{1}{3}$ C) $\dfrac{1}{4}$ D) $\dfrac{1}{5}$ E) $\dfrac{1}{6}$

3. $\lim\limits_{m \to x} \dfrac{m^3 - x^3}{m - x} = ?$

A) $2x^2$ B) $3x^2$ C) $4x^2$ D) $6x^2$ E) x^2

4. $\lim\limits_{\alpha \to 1} \dfrac{\sqrt[4]{\alpha} - 1}{\sqrt[3]{\alpha} - 1} = ?$ 📷

A) $-\dfrac{4}{3}$　　B) $-\dfrac{3}{4}$　　C) 1　　D) $\dfrac{3}{4}$　　E) $\dfrac{4}{3}$

5. $\lim\limits_{x \to 1} (4x - \ln x) = ?$

A) 1　　B) 2　　C) 3　　D) 4　　E) 5

6. $\lim\limits_{x \to c} \dfrac{x\sqrt{x} - c\sqrt{c}}{\sqrt{x} - \sqrt{c}} = ?$

A) $\dfrac{c}{3}$　　B) $\dfrac{c}{2}$　　C) c　　D) $2c$　　E) $3c$

7. $\lim\limits_{x \to 2} \dfrac{\sqrt[3]{x + 6} - 2}{x - 2} = ?$

A) $-\dfrac{1}{1}$　　B) $-\dfrac{1}{8}$　　C) 0　　D) $\dfrac{1}{12}$　　E) $\dfrac{1}{4}$

8. $\lim\limits_{x \to 2} \dfrac{\sqrt[3]{x} - \sqrt[3]{2}}{x - 2} = ?$

A) $\dfrac{1}{2}$　　B) $\dfrac{1}{3 \cdot \sqrt[3]{6}}$　　C) $\dfrac{1}{3 \cdot \sqrt[3]{4}}$　　D) $\dfrac{1}{3 \cdot \sqrt[3]{2}}$　　E) $\dfrac{1}{6}$ 📷

9. $\lim\limits_{x\to 2}\dfrac{x-\sqrt{2x}}{\sqrt{2x+5}-3}=\ ?$

A) $\dfrac{1}{2}$ B) 1   C) $\dfrac{3}{2}$ D) 2   E) $\dfrac{5}{2}$

10. $\lim\limits_{x\to 1}\dfrac{1-\sqrt[p]{x}}{1-\sqrt[q]{x}}=\ ?$

A) $-\dfrac{p}{q}$ B) $-\dfrac{q}{p}$ C) $\dfrac{p}{q}$ D) $\dfrac{q}{p}$ E) $p\cdot q$

11. $\lim\limits_{x\to 2}\dfrac{\sqrt{x^2-3x+6}-\sqrt{x^2-2x+4}}{4x-8}=\ ?$

A) $-\dfrac{1}{4}$ B) $-\dfrac{1}{8}$ C) $-\dfrac{1}{16}$ D) $\dfrac{1}{8}$ E) $\dfrac{1}{16}$

12. $\lim\limits_{x\to 2}\dfrac{8-x^3}{x^2-2x+3}=\ ?$

A) -2 B) -1 C) 0   D) 1   E) 2

13. $\lim\limits_{x\to 9}\dfrac{\sqrt{x}-3}{\sqrt[4]{x}-\sqrt{3}}=\ ?$

A) $2\sqrt{3}$ B) $\sqrt{15}$ C) 4   D) $3\sqrt{2}$ E) $3\sqrt{3}$

14. $\lim\limits_{\frac{a}{b}\to 1} \dfrac{4a + 3b}{2a + b} = ?$

A)$\dfrac{2}{7}$ B)$\dfrac{3}{7}$ C) 2   D)$\dfrac{7}{3}$ E)$\dfrac{7}{2}$

15. $\lim\limits_{x\to 0} \dfrac{\sqrt[3]{x + 27} - 3}{\sqrt{x + 64} - 8} = ?$

A) $-\dfrac{27}{22}$ B) $-\dfrac{32}{27}$ C)$\dfrac{27}{32}$ D)$\dfrac{32}{27}$ E)$\dfrac{16}{27}$

16. $\lim\limits_{x\to 5} \left(\dfrac{1}{x - 5} - \dfrac{3x - 8}{x^2 - 3x - 10}\right) = ?$

A) $-\dfrac{5}{7}$ B) $-\dfrac{2}{7}$ C) 0   D)$\dfrac{2}{7}$ E)$\dfrac{5}{7}$

17. $\lim\limits_{x\to 1} \dfrac{ax - \sqrt{x + 3}}{x^2 - 1} = b, b \in R \Rightarrow a = ?$

A) 2   B) 3   C) 4   D) 5   E) 6

18. $\lim\limits_{x\to\frac{\pi}{2}} \dfrac{\pi - 2x}{\cos x} = ?$

A) – 2   B) – 1   C) 0   D) 1   E) 2

19. $\lim\limits_{y \to x} \dfrac{\sin^2 y - \sin^2 x}{y^2 - x^2} = ?$ 🔟

A) $\sin x$ B) $\dfrac{\sin x}{2x}$ C) $\dfrac{\sin 2x}{x}$ D) $\dfrac{\sin x}{x}$ E) $\dfrac{\sin 2x}{2x}$

20. $\lim\limits_{x \to e} \dfrac{\ln x - 1}{x^2 - e^2} = ?$

A) $\dfrac{1}{e}$ B) $\dfrac{1}{e^2}$ C) $\dfrac{1}{2e}$ D) $\dfrac{1}{2e^2}$ E) $\dfrac{1}{4e^2}$

21. $\lim\limits_{x \to 1} \tan\left(\dfrac{\pi}{2}x\right) \cdot (x - 1) = ?$ 🔟

A) $\dfrac{2}{\pi}$ B) $-\dfrac{\pi}{2}$ C) $-\pi$ D) π E) 2π

Answers					
1. D	2. B	3. B	4. D	5. D	6. E
7. D	8. C	9. C	10. D	11. A	12. C
13. A	14. D	15. E	16. B	17. A	18. E
19. E	20. D	21. A			

Test 3

1.$\lim\limits_{x\to\pi} (\cos(\sin x)) = ?$

A) 1   B) 0   C) – 1   D) $-\dfrac{1}{2}$ E) $-\dfrac{1}{3}$

2. $\lim\limits_{x\to -1} \dfrac{(2+3x)^2 - (x+2)^2}{x^3 - x} = ?$

A) – 4   B) – 2   C) 0   D) 2   E) 4

3. $\lim\limits_{x\to -1} \dfrac{3x^3 + 3x^2 + x + 1}{x^2 - 1} = ?$

A) – 3   B) – 2   C) – 1   D) 0   E) 1

4.$\lim\limits_{x\to 3} \dfrac{x^2 - 9}{\sqrt{x+6} - 3} = ?$

A) 4   B) 9   C) 16   D) 36   E) 39

5.$\lim\limits_{x\to \frac{\pi}{2}} \dfrac{\sin 2x - \cos x}{x - \dfrac{\pi}{2}} = ?$

A) – 1   B) – 2   C) – 3   D) 1   E) 2

$6.\lim\limits_{x\to1}\left(\dfrac{x+m}{\sqrt{3+x}-2}\right)=k,\,k\in R\Rightarrow m=\,?$

A) – 1   B) – 2   C) – 3   D) – 4   E) – 5

$7.\lim\limits_{x\to\frac{\pi}{2}}\left(\dfrac{1+\cos 2x}{1-\sin x}\right)=\,?$

A) 0   B) 2   C) – 2   D) 4   E) – 4

$8.\lim\limits_{x\to0}\left(\dfrac{x^2+\sin^2 x}{\tan^2 x}\right)=\,?$

A) 3   B) $\dfrac{3}{2}$ C) 2   D) 1   E) $\dfrac{5}{2}$

$9.\lim\limits_{x\to2}\left(\dfrac{x^3-ax^2+3x-2}{x^2-4}\right)=k,\,k\in R\Rightarrow k=\,?$

A) 3   B) $\dfrac{3}{4}$ C) $\dfrac{1}{2}$ D) 2   E) $-\dfrac{3}{2}$

$10.\lim\limits_{x\to1}\left(\dfrac{\sqrt{x+3}-\sqrt{3x+1}}{x^2-3x+2}\right)=\,?$

A) 1   B) $\dfrac{1}{2}$ C) $\dfrac{1}{3}$ D) 0   E) – 1

44

11. $\lim\limits_{x \to \frac{\pi}{4}} \dfrac{\sqrt{2}\cos x - 1}{1 - \tan^2 x} = $?

A) -2 B) $-\dfrac{1}{2}$ C) 0 D) $\dfrac{1}{4}$ E) $\dfrac{1}{2}$

12. $\lim\limits_{x \to 0} \dfrac{5e^x + 5e^{-x} - 10}{4x^2} = $?

A) $-\dfrac{5}{2}$ B) -1 C) 0 D) $\dfrac{5}{4}$ E) 1

13. $\lim\limits_{x \to 3} \left(\dfrac{1}{x-3} - \dfrac{6}{x^2-9} \right) = $?

A) $\dfrac{1}{4}$ B) 4 C) 3 D) $\dfrac{1}{3}$ E) $\dfrac{1}{6}$

14. $\lim\limits_{x \to \infty} \dfrac{(b-1)x^3 + (2a-6)x^2 + x - 1}{ax^2 - 2x + 5} = -1 \Rightarrow a+b = $?

A) 3 B) 2 C) 1 D) 0 E) ∞

15. $\lim\limits_{x \to \infty} \left(\dfrac{\sin x}{x} + \dfrac{x^2+3x}{2x^2-1} \right) = $?

45

A) $\dfrac{3}{2}$ B) $\dfrac{1}{2}$ C) 1   D) 2   E) ∞

16. $\displaystyle \lim_{x\to\infty} \left(\sqrt{x^2-1} - \sqrt{x^2+3x+1}\right) = ?$

A) -2 B) $-\dfrac{5}{2}$ C) $-\dfrac{3}{2}$ D) 0   E) -1

17. $\displaystyle \lim_{x\to\infty} \left(\dfrac{x}{x+2}\right)^x = ?$

A) e^{-2} B) e^2 C) e D) 0   E) -1

18. $\displaystyle \lim_{x\to\infty} \left(\dfrac{x-2}{x+2}\right)^{x+2} = ?$

A) e^4 B) e^{-4} C) e^2 D) e^{-2} E) e

19. $\displaystyle \lim_{x\to\frac{\pi}{2}} \dfrac{sin\left(x - \dfrac{\pi}{2}\right)}{cos\,x} = k,\ k \in R$

$\Rightarrow \displaystyle \lim_{x\to k} \left(3^{x+1} + e^{-x}\right) = ?$

A) e B) $e+3$ C) $e+1$ D) $e+2$ E) $2e$

46

20. $\lim\limits_{x \to 0} \dfrac{ln(3x + 1)}{2x} = ?$

A)$\dfrac{3}{2}$ B)$\dfrac{2}{3}$ C)$\dfrac{1}{2}$ D) 3   E) 12

21. $\lim\limits_{x \to 2} [(x - 2) \cdot ln(x - 2)] = ?$

A) – 2   B) – 1   C) 0   D) 1   E) 2

22. $\lim\limits_{x \to 0} \left(\dfrac{1}{sin\ x} - \dfrac{1}{x} \right) = ?$

A) 0   B) 1   C) 2   D) 3   E) 4

23. $\lim\limits_{x \to 0} \dfrac{\sqrt[4]{1 + 2x} - 1}{x} = ?$

A) – 2   B) $-\dfrac{1}{2}$ C) 0   D)$\dfrac{1}{2}$ E) 2

Answers					
1. C	2. A	3. B	4. D	5. A	6. A
7. D	8. C	9. B	10. B	11. D	12. D
13. E	14. A	15. B	16. C	17. A	18. B
19. C	20. A	21. C	22. A	23. D	

Test 4

1. $\lim\limits_{x \to 3} [4 \cdot (3x - 2) \cdot (x + 2)] = ?$

A) 100   B) 120   C) 135   D) 140   E) 150

2. $\lim\limits_{x \to -6} [(x + 4)^{100} \cdot (x + 2)] = ?$

A) – 3   B) – 2   C) – 1   D) 2   E) 3

3. $\lim\limits_{x \to 1} \dfrac{x^2 - 2x + 1}{(x^3 - 1)^2} = ?$

A) $\dfrac{1}{9}$ B) $\dfrac{1}{3}$ C) 1   D) 3   E) 9

4. $\lim\limits_{a \to 0} \dfrac{(x + a)^3 - x^3}{a} = ?$

48

A) $\dfrac{x}{3}$ B) $3x$ C) $3x^2$ D) $6x$ E) $9x^2$

5. $\displaystyle\lim_{x\to 2}\left(\dfrac{1}{(x-2)}-\dfrac{1}{x^2-3x+2}\right)=$?

A) -2 B) -1 C) 0 D) 1 E) 2

6. $\displaystyle\lim_{x\to 16}\dfrac{\sqrt{x}-4}{\sqrt[4]{x}-2}=$?

A) -16 B) -4 C) 0 D) 4 E) 16

7. $\displaystyle\lim_{x\to 2}\dfrac{\sqrt{x+2}-x}{3-\sqrt{4x+1}}=$?

A) 1 B) $\dfrac{9}{8}$ C) $\dfrac{5}{4}$ D) $\dfrac{11}{8}$ E) $\dfrac{4}{3}$

8. $\displaystyle\lim_{x\to 0}\dfrac{\sqrt{x^2+m}-1}{x}=n,\ m,n\in R\Rightarrow n=$?

A) -2 B) -1 C) 0 D) 1 E) 2

9. $\displaystyle\lim_{x\to\infty}\left(\dfrac{x^2}{2x+5}-\dfrac{x}{2}\right)=$?

A) $-\dfrac{1}{4}$ B) $-\dfrac{1}{2}$ C) $-\dfrac{3}{4}$ D) -1 E) $-\dfrac{5}{4}$

10. $\lim\limits_{x\to\infty} \left(\dfrac{x^3}{x^2+2} - x \right) = ?$

A) 0   B) 1   C) 2   D) 3   E) 4

11. $\lim\limits_{x\to -\infty} \dfrac{4}{1-3^{x/1-x}} = ?$

A) 4   B) 6   C) 8   D) 10   E) 12

12. $\lim\limits_{x\to -\infty} \dfrac{3\cdot 2^x + 7}{4 - 2^{x+1}} = ?$

A) $\dfrac{3}{2}$ B) $\dfrac{13}{5}$ C) $\dfrac{7}{4}$ D) $\dfrac{21}{8}$ E) 5

13. $\lim\limits_{x\to\infty} \left(\sqrt{x^2-8x} - x \right) = ?$

A) -8 B) -4 C) -2 D) 4   E) 8

14. $\lim\limits_{x\to m} \dfrac{\sin m - \sin x}{\cos x - \cos m} = ?$

A) $\cot m$ B) $-\cos m$ C) $-\dfrac{1}{\cos m}$ D) $-\dfrac{1}{\sin m}$ E) $-\sin m$

50

15. $\lim\limits_{x \to \frac{\pi}{4}} \dfrac{sin\left(x - \dfrac{\pi}{4}\right)}{cos\left(x + \dfrac{\pi}{4}\right)} = ?$

A) $-\dfrac{1}{3}$ B) -1 C) $\dfrac{1}{4}$ D) $\dfrac{1}{2}$ E) $\dfrac{\sqrt{2}}{2}$

16. $\lim\limits_{x \to \frac{\pi}{4}} \dfrac{sin\, x - cos\, x}{cot\, x - 1} = ?$

A) $-2\sqrt{2}$ B) $-\sqrt{2}$ C) $-\dfrac{\sqrt{2}}{2}$ D) $\dfrac{\sqrt{2}}{2}$ E) $\sqrt{2}$

17. $\lim\limits_{x \to \frac{\pi}{3}} \dfrac{1 - 2cos\, x}{sin\left(x - \dfrac{\pi}{3}\right)} = ?$

A) $\dfrac{\sqrt{2}}{2}$ B) 1 C) $\sqrt{2}$ D) $\dfrac{3}{2}$ E) $\sqrt{3}$

18. $\lim\limits_{x \to 0} \dfrac{\sqrt[3]{1 + x^2} - \sqrt[4]{1 - 2x}}{x} = ?$

A) $\dfrac{1}{4}$ B) $\dfrac{1}{2}$ C) $\dfrac{3}{4}$ D) 1 E) 2

19. $\lim\limits_{x \to 2} \dfrac{x^3 - 3x^2 + x + 2}{x^4 - 4x - 8} = ?$

A)$\dfrac{1}{64}$ B)$\dfrac{1}{32}$ C)$\dfrac{1}{28}$ D)$\dfrac{1}{12}$ E)$\dfrac{1}{6}$

20. $\lim\limits_{x \to 1} \left(\dfrac{1}{\ln x} - \dfrac{1}{x - 1} \right) = ?$

A) 4   B) 2   C) 1   D)$\dfrac{1}{2}$ E)$\dfrac{1}{4}$

21. $\lim\limits_{x \to 2} \dfrac{3 - \sqrt{5x - 1}}{\sqrt{x + 2} - x} = ?$

A)$\dfrac{8}{3}$ B) 2   C)$\dfrac{5}{3}$ D)$\dfrac{4}{3}$ E)$\dfrac{10}{9}$

Answers					
1. D	2. A	3. A	4. C	5. D	6. D
7. B	8. C	9. E	10. A	11. B	12. C
13. B	14. A	15. B	16. C	17. E	18. B
19. C	20. C	21. E			

Test 5

1. $\lim\limits_{x \to 0} \dfrac{\sin^3 \dfrac{x}{2}}{x^3} = ?$

A) 16   B) 8   C) 4   D)$\dfrac{1}{8}$ E)$\dfrac{1}{16}$

2. $\lim\limits_{x \to 8} \dfrac{\sqrt[3]{x} - 2}{\sqrt{x - 2\sqrt{2}}} = ?$

A)$\dfrac{\sqrt{2}}{3}$ B)$\dfrac{\sqrt{2}}{6}$ C)$\dfrac{\sqrt{2}}{8}$ D)$\dfrac{1}{2}$ E)$\dfrac{1}{4}$

3. $\lim\limits_{x \to \pi} \dfrac{\pi \cos 2x - x}{1 + \cos x} = ?$

A) – 1   B) – 2   C) 0   D) 1   E) 2

4. $\lim\limits_{x \to 2} \dfrac{x - 2}{x^3 - 8} = ?$

A)$\dfrac{1}{3}$ B)$\dfrac{1}{6}$ C)$\dfrac{1}{8}$ D)$\dfrac{1}{12}$ E)$\dfrac{3}{16}$

5. $\lim\limits_{a \to x} \dfrac{a^3 - x^3}{a^2 - x^2} = ?$

A)$\dfrac{3x^2}{2}$ B)$\dfrac{3x}{2}$ C)$\dfrac{3}{2x}$ D)$\dfrac{x}{3}$ E)$\dfrac{x}{2}$

6. $\lim\limits_{x \to 1} \dfrac{\tan\dfrac{\pi}{4}x - \cos 2\pi}{x - 1} = ?$

A) π B)$\dfrac{\pi}{2}$ C)$\dfrac{\pi}{4}$ D)$\dfrac{3\pi}{4}$ E)$\dfrac{\pi}{6}$

7. $\lim\limits_{x \to 0} \dfrac{\sin 4x}{x} = ?$

A) 1   B) 2   C) 4   D)$\dfrac{1}{4}$ E)$\dfrac{1}{8}$

8. $\lim\limits_{x \to 0} \dfrac{1 - \cos x}{\sin^2 x} = ?$

A) 1   B)$\dfrac{1}{2}$ C)$\dfrac{1}{4}$ D)$\dfrac{3}{2}$ E)$\dfrac{3}{4}$

9. $\lim\limits_{x \to \infty} \left(\dfrac{x + 7}{x + 3}\right)^x = ?$

A)$\dfrac{1}{4}$ B) e^2 C) e^3 D) e^4 E)$\dfrac{1}{4}e$

10. $\lim\limits_{x\to\infty}\left(\dfrac{x+5}{x+1}\right)^{3x+2} = ?$

A) e B) e^3 C) e^4 D) e^6 E) e^{12}

11. $\lim\limits_{x\to 2}\dfrac{x-2}{\sqrt{2x-2}} = ?$

A) 1   B) 2   C)$\dfrac{1}{2}$ D)$\dfrac{1}{4}$ E)$\dfrac{1}{8}$

12. $\lim\limits_{x\to e}\dfrac{\ln x-1}{x-e} = ?$

A) 1   B)$\dfrac{1}{2}$ C) e D)$\dfrac{1}{e}$ E)$\dfrac{1}{e^2}$

13. $\lim\limits_{x\to 0}\dfrac{\ln(1+4x)}{\sin 4x} = ?$

A) 16   B) 8   C) 4   D) 2   E) 1

14. $\lim\limits_{x\to 1}\dfrac{\sin \pi x}{1-x^2} = ?$

A) $-\pi$ B) $-\dfrac{\pi}{2}$ C) $\dfrac{\pi}{2}$ D) π E) 2π

15. $\lim\limits_{x\to 0}\dfrac{\cos x - \cos^3 x}{x^2} = ?$

A) -2 B) -1 C) 0 D) 1 E) 2

16. $\lim\limits_{x\to\frac{\pi}{3}}\dfrac{3x - \pi}{\sin\left(x - \dfrac{\pi}{3}\right)} = ?$

A) -9 B) -3 C) 0 D) 3 E) 9

17. $\lim\limits_{x\to\pi}\dfrac{2x - 2\pi}{\tan x - \sin x} = ?$

A) 2 B) 1 C) 0 D) -1 E) -2

18. $\lim\limits_{x\to\frac{\pi}{2}}\dfrac{\sin(\cos x)}{\cos^2\dfrac{x}{2} - \sin^2\dfrac{x}{2}} = ?$

A) $-\sqrt{2}$ B) $\dfrac{-\sqrt{2}}{2}$ C) 1 D) $\dfrac{\sqrt{2}}{2}$ E) $\sqrt{2}$

19. $\lim\limits_{x \to 0} \dfrac{\sin 5x - \sin 2x}{\sin x} = ?$

A) 1   B) $\dfrac{3}{2}$ C) 2   D) $\dfrac{5}{2}$ E) 3

20. $\lim\limits_{x \to \infty} \left(\dfrac{4x^2 + 6}{2x^2 - 7x + 1} + 5^{\frac{1}{x}} \right) = ?$

A) – 2   B) – 1   C) 0   D) 1   E) 3

21. $\lim\limits_{x \to \frac{\pi}{4}} \dfrac{\cos 2x}{\cos x - \sin x} = ?$

A) 1   B) $\sqrt{2}$ C) $\sqrt{3}$ D) 2   E) $2\sqrt{2}$

22. $\lim\limits_{0 \to x} \dfrac{\tan x - \tan\theta}{\tan(\theta - x)} = ?$

A) $- \sec^2 x$ B) $\sec^2 x$ C) $\operatorname{cosec}^2 x$ D) $- \operatorname{cosec}^2 x$ E) $\tan x$

23. $\lim\limits_{x \to 3} \dfrac{x^3 - 3x^2 + x - 3\text{🔟}}{x^2 - x - 6} = ?$

A) – 3   B) – 2   C) 0   D) 2   E) 3

Answers					
1. D	2. A	3. A	4. D	5. B	6. B

7. C	8. B	9. D	10. E	11. B	12. D
13. E	14. C	15. D	16. D	17. B	18. C
19. E	20. E	21. B	22. B	23. D	

Chapter   **Limit**

Test 6

1. $\lim\limits_{t \to \infty} \left(\dfrac{t^3 - 2}{t^3 + 1}\right)^{t^3 + 3} = ?$

A) 1   B) 2   C) e   D) e^{-2} E) ∞

2. $\lim\limits_{x \to -\infty} (\sqrt{x^2 - 4x + 6} - \sqrt{x^2 + ax + 3}) = 6 \Rightarrow a = ?$

A) -16 B) -12 C) -8 D) 8   E) 16

3. $\lim\limits_{x \to 6} \dfrac{\sqrt{5x + 6} - 6}{\sqrt{x + 3} - 3} = ?$

A) $\dfrac{5}{2}$ B) 3   C) $\dfrac{7}{2}$ D) 4   E) $\dfrac{9}{2}$

4. $\lim\limits_{x\to2} (2 - x)\cdot tan\left(\dfrac{\pi}{4}x\right) = ?$ 🔟

A)$\dfrac{\pi}{2}$ B)$\dfrac{2}{3}$ C)$\dfrac{2\pi}{3}$ D)$\dfrac{2}{\pi}$ E)$\dfrac{4}{\pi}$

5. $\lim\limits_{x\to0} \dfrac{\sqrt[3]{x + 1^{10}} - 1}{\sqrt{x + 1} - 1} = ?$

A) 2   B)$\dfrac{3}{2}$ C) 1   D)$\dfrac{2}{3}$ E)$\dfrac{3}{4}$

6. $\lim\limits_{x\to0} \dfrac{tan(x^2) + tan^2 x}{x^2} = ?$

A) 1   B)$\dfrac{3}{2}$ C) 2   D)$\dfrac{5}{2}$ E) 3

7. $\lim\limits_{x\to-4} \left(\dfrac{1}{x + 4} - \dfrac{8}{16 - x^2}\right) = ?$

A)$\dfrac{1}{2}$ B)$\dfrac{1}{4}$ C)$\dfrac{1}{8}$ D) $-\dfrac{1}{4}$ E) $-\dfrac{1}{8}$

8. $\lim\limits_{x\to\frac{\pi}{4}} tan(2x)(tan\ x - 1) = ?$

A) -1 B) $-\dfrac{1}{2}$ C) 0   D) 1   E)$\dfrac{1}{2}$

59

9. $\lim\limits_{x \to 0} \dfrac{\sqrt{x + 16} - 4}{\sin 8x} = ?$

A)$\dfrac{1}{4}$ B)$\dfrac{1}{8}$ C)$\dfrac{1}{16}$ D)$\dfrac{1}{32}$ E)$\dfrac{1}{64}$

10. $\lim\limits_{x \to 0} \cot x(\operatorname{cosec} x - \cot x) = ?$

A) 1   B)$\dfrac{1}{2}$ C)$\dfrac{1}{4}$ D)$\dfrac{1}{8}$ E)$\dfrac{3}{4}$

11. $\lim\limits_{x \to 4} \dfrac{x^2 - 16}{\sqrt{x - 1} - \sqrt{3}} = ?$

A) $2\sqrt{3}$ B) $4\sqrt{3}$ C) $8\sqrt{3}$ D) $12\sqrt{3}$ E) $16\sqrt{3}$

12. $\lim\limits_{x \to 25} \dfrac{\sqrt{x} - 5}{x - 25} = ?$

A)$\dfrac{1}{5}$ B)$\dfrac{1}{10}$ C)$\dfrac{1}{15}$ D)$\dfrac{1}{20}$ E)$\dfrac{1}{25}$

13. $\lim\limits_{x \to 2} \left(\dfrac{1}{x - 2} - \dfrac{12}{x^3 - 8} \right) = ?$

A)$\dfrac{3}{2}$ B) 1   C)$\dfrac{1}{2}$ D)$\dfrac{1}{4}$ E)$\dfrac{1}{8}$

14. $\lim\limits_{x\to\infty} \left(\sqrt{x^2 + 2x} - x\right) = ?$

A) 1   B)$\dfrac{3}{2}$ C) 2   D)$\dfrac{5}{2}$ E) 4

15. $\lim\limits_{x\to 2} \dfrac{\sqrt{4 - x^2}}{\sqrt{6 - 5x + x^2}} = ?$

A) – 2   B) – 1   C) 0   D) 1   E) 2

16. $\lim\limits_{x\to 2} \sqrt{\dfrac{x^4 - 16}{x^3 - 8}} = ?$

A)$\dfrac{\sqrt{6}}{3}$ B)$\dfrac{2\sqrt{6}}{3}$ C)$\dfrac{2\sqrt{3}}{3}$ D)$\dfrac{\sqrt{6}}{4}$ E) $2\sqrt{2}$

17. $\lim\limits_{x\to\frac{\pi}{2}} \dfrac{\sin x - 1}{\sin x^2 - 1} = ?$

A) 0   B)$\dfrac{1}{2}$ C) 1   D)$\dfrac{3}{2}$ E) 3

18. $\lim\limits_{x\to 4} \dfrac{\sqrt{x} - 2}{x^2 - 16} = ?$

$$A) \frac{1}{4} \quad B) \frac{1}{8} \quad C) \frac{1}{16} \quad D) \frac{1}{32} \quad E) \frac{1}{64}$$

19. $\lim\limits_{x \to \frac{\pi}{2}} \dfrac{4 - 4\sin x}{\sin 4x} = ?$

A) 2   B) 1   C) 0   D) – 1   E) – 2

20. $\lim\limits_{x \to 0} \dfrac{8x + \sin(6x)}{x^2 - 4x + \sin(12x)} = ?$

$$A) \frac{3}{2} \quad B) \frac{7}{2} \quad C) \frac{7}{4} \quad D) \frac{1}{2} \quad E) 1$$

21. $\lim\limits_{x \to \infty} \left(\sqrt{x^2 + 7x} - \sqrt{x^2 + 2x} \right) = ?$

$$A) 5 \quad B) 3 \quad C) \frac{5}{2} \quad D) 2 \quad E) \frac{1}{2}$$

22. $\lim\limits_{x \to -\infty} \left(\sqrt{4x^2 + 6x} - \sqrt{4x^2 + 9} \right) = ?$

$$A) -\frac{3}{2} \quad B) -\frac{1}{2} \quad C) 1 \quad D) \frac{3}{2} \quad E) \frac{5}{2}$$

Answers					
1. E	2. D	3. A	4. E	5. D	6. C

7. E	8. A	9. E	10. B	11. E	12. B
13. C	14. A	15. E	16. B	17. B	18. D
19. C	20. C	21. C	22. A		

Chapter   **Limit**

Test 7

1. $\lim\limits_{x \to -2} (4 - x^2) \cdot tan\left(\dfrac{\pi x}{4}\right) = $?

A)$\dfrac{\pi}{2}$ B)$\dfrac{\pi}{4}$ C)$\dfrac{\pi}{6}$ D)$\dfrac{\pi}{8}$ E)$\dfrac{\pi}{12}$

2. $\lim\limits_{x \to 1} \dfrac{tan(27 - x^3)}{x^4 - 81} = $?

A) $-\dfrac{1}{4}$ B) $-\dfrac{1}{2}$ C) 0 D)$\dfrac{1}{4}$ E)$\dfrac{3}{8}$

3. $\lim\limits_{x \to \frac{\pi}{2}} \dfrac{(1 + \sin x)\cdot \cos x}{\pi + 2x} = ?$

A) 1   B) $\dfrac{1}{2}$ C) 0   D) $-\dfrac{1}{2}$ E) -1

4. $\lim\limits_{x \to 2} \dfrac{\sin\left(\dfrac{\pi x}{4} - \dfrac{\pi}{2}\right)}{\ln(3x - 5)} = ?$

A) $\dfrac{\pi}{2}$ B) $\dfrac{\pi}{4}$ C) $\dfrac{\pi}{6}$ D) $\dfrac{3\pi}{4}$ E) $\dfrac{\pi}{12}$

5. $\lim\limits_{x \to 0} \dfrac{\sin(e^{6x} - 1)}{\sin(e^{2x} - 1)} = ?$

A) $\dfrac{1}{2}$ B) 1   C) $\dfrac{3}{2}$ D) 3   E) 6

6. $\lim\limits_{x \to 1} \sec\left(\dfrac{\pi}{2}x\right)\cdot\left(\arctan x - \dfrac{\pi}{4}\right) = ?$

A) -1 B) $-\dfrac{1}{2}$ C) $-\dfrac{1}{\pi}$ D) $\dfrac{2}{\pi}$ E) $\dfrac{\pi}{2}$

64

7. $\lim\limits_{x\to 0} \dfrac{2\sin(5x)}{3x} = ?$

A)$\dfrac{12}{5}$ B) 3   C)$\dfrac{10}{3}$ D)$\dfrac{7}{2}$ E) 4

8. $\lim\limits_{x\to 0} \sin(5x)\cdot\cot(3x) = ?$

A)$\dfrac{4}{3}$ B)$\dfrac{3}{2}$ C)$\dfrac{5}{3}$ D)$\dfrac{7}{3}$ E)$\dfrac{5}{2}$

9. $\lim\limits_{x\to 0} x\cdot\operatorname{cosec}^2\sqrt{2x} = ?$

A)$\dfrac{3}{2}$ B) 1   C)$\dfrac{3}{4}$ D)$\dfrac{2}{3}$ E)$\dfrac{1}{2}$

10. $\lim\limits_{x\to 0} \dfrac{\sin 2x}{2x^2 + x} = ?$

A)$\dfrac{5}{2}$ B) 2   C)$\dfrac{3}{2}$ D)$\dfrac{2}{3}$ E) 0

11. $\lim\limits_{x\to 0} x\cdot\cot(2x) = ?$

A)$\dfrac{3}{2}$ B) 1   C)$\dfrac{2}{3}$ D)$\dfrac{1}{2}$ E)$\dfrac{1}{4}$

$12.\lim\limits_{x\to 0}\dfrac{x^2+4x}{sin(3x)} = ?$

A)$\dfrac{3}{2}$ B)$\dfrac{4}{3}$ C)$\dfrac{2}{3}$ D)$\dfrac{1}{2}$ E) 0

13. $\lim\limits_{x\to 0} tan\,3x \cdot cos\,6x = ?$

A) 0   B)$\dfrac{2}{3}$ C)$\dfrac{3}{2}$ D) 3   E)$\dfrac{9}{2}$

$14.\lim\limits_{x\to\frac{\pi}{2}}\dfrac{4x-2\pi}{cos\,x} = ?$

A) 4   B) 2   C) 0   D) – 2   E) – 4

15. $f(x) = \dfrac{\sqrt[3]{x}+2}{x+8} \Rightarrow \lim\limits_{x\to -6} f(x) = ?$

A) 1   B)$\dfrac{1}{2}$ C)$\dfrac{1}{4}$ D)$\dfrac{1}{8}$ E)$\dfrac{1}{12}$

$16.\lim\limits_{x\to 0}\dfrac{sin\,2x\cdot tan\,2x}{1-cos\,4x} = ?$

A)$\dfrac{1}{8}$ B)$\dfrac{1}{4}$ C)$\dfrac{1}{2}$ D) 4   E) 2

17. $\lim\limits_{x \to 1} \dfrac{x^6 - x}{x^2 + 7x - 8} = ?$

A) $-\dfrac{2}{3}$ B) $-\dfrac{5}{6}$ C) $\dfrac{4}{3}$ D) $\dfrac{2}{3}$ E) $\dfrac{4}{9}$

18. $\lim\limits_{x \to 8} \dfrac{x - 8}{\sqrt[3]{x - 16} + 12} = ?$

A) $\dfrac{5}{6}$ B) $\dfrac{4}{3}$ C) 6   D) 12   E) 18

19. $\lim\limits_{x \to 0} \dfrac{2x + \sin 4x}{\sin 8x} = ?$

A) $\dfrac{1}{2}$ B) $\dfrac{2}{3}$ C) $\dfrac{3}{4}$ D) 1   E) $\dfrac{2}{3}$

20. $\lim\limits_{x \to \frac{\pi}{2}} \dfrac{\cos^2 x}{1 - \sin^3 x} = ?$

A) $\dfrac{1}{2}$ B) $\dfrac{2}{3}$ C) 1   D) $\dfrac{3}{2}$ E) $\dfrac{5}{2}$

21. $\lim\limits_{x \to \frac{\pi}{4}} \dfrac{\sin x - \cos x}{1 - \tan x} = ?$

A) $-\dfrac{\sqrt{2}}{2}$ B) 0   C)$\dfrac{\sqrt{2}}{2}$ D) 1   E)$\dfrac{3}{2}$

22. $\lim\limits_{x\to 2} \dfrac{4x^2 - 2x + m - 2}{x^2 - 4} \in R \Rightarrow m = ?$

A)$\dfrac{5}{2}$ B) 3   C)$\dfrac{7}{2}$ D) 4   E)$\dfrac{9}{2}$

23. $\lim\limits_{x\to 0} \dfrac{\sqrt{x + 9} - 3}{\sqrt{x + 16} - 4} = ?$

A) 1   B)$\dfrac{4}{3}$ C)$\dfrac{5}{2}$ D) 3   E) 4

Answers					
1. D	2. A	3. E	4. E	5. D	6. C
7. C	8. C	9. E	10. B	11. D	12. B
13. A	14. E	15. E	16. C	17. E	18. D
19. C	20. B	21. A	22. C	23. B